AMAZING ANIMAL CLASSES

Brilliant Birds

by Betsy Rathburn
Illustrated by Natalya Karpova

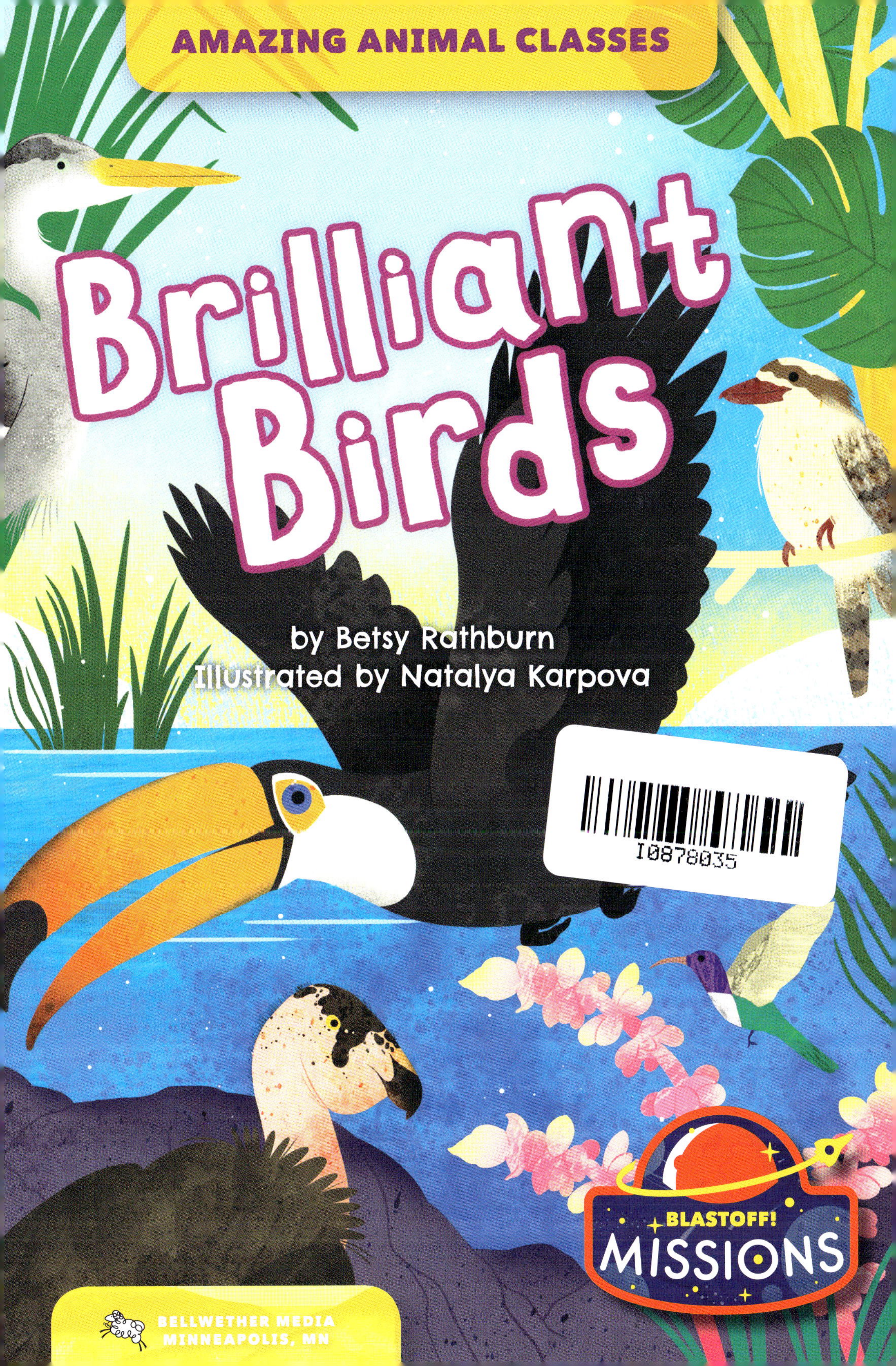

BLASTOFF! MISSIONS

BELLWETHER MEDIA
MINNEAPOLIS, MN

Blastoff! Missions takes you on a learning adventure! Colorful illustrations and exciting narratives highlight cool facts about our world and beyond. Read the mission goals and follow the narrative to gain knowledge, build reading skills, and have fun!

Traditional Nonfiction

Narrative Nonfiction

Blastoff! Universe

MISSION GOALS

- FIND YOUR SIGHT WORDS IN THE BOOK.
- IDENTIFY THE TRAITS OF BIRDS.
- THINK OF QUESTIONS TO ASK WHILE YOU READ.

This edition first published in 2023 by Bellwether Media, Inc.

Library of Congress Cataloging-in-Publication Data

Names: Rathburn, Betsy, author.
Title: Brilliant birds / by Betsy Rathburn.
Description: Minneapolis, MN : Bellwether Media, Inc., 2023. | Series: Blastoff! missions: amazing animal classes | Includes bibliographical references and index. | Audience: Ages 5-8 | Audience: Grades 2-3 | Summary: "Vibrant illustrations accompany information about birds. The narrative nonfiction text is intended for students in kindergarten through third grade"-- Provided by publisher.
Identifiers: LCCN 2022020214 (print) | LCCN 2022020215 (ebook) | ISBN 9781644876480 (library binding) | ISBN 9781648348327 (paperback) | ISBN 9781648346941 (ebook)
Subjects: LCSH: Birds--Juvenile literature.
Classification: LCC QL676.2 .R345 2023 (print) | LCC QL676.2 (ebook) | DDC 598--dc23/eng/20220504
LC record available at https://lccn.loc.gov/2022020214
LC ebook record available at https://lccn.loc.gov/2022020215

Editor: Christina Leaf Designer: Andrea Schneider

Printed in the United States of America, North Mankato, MN.

This is **Blastoff Jimmy**! He is here to help you on your mission and share fun facts along the way!

Table of Contents

Let's Go Bird-watching!

Grab your **binoculars**! Let's go on a bird-watching journey!

These **warm-blooded** animals have backbones and lay eggs. Feathers and **hollow** bones help many birds fly.

High Fliers

Look up! See that California condor flying over North America's Mojave Desert? It is looking for **carrion** to eat.

These Costa's hummingbirds want food, too. **Nectar** is on their menu!

JIMMY SAYS

California condors are the largest birds in North America!

Watch for the bright feathers of scarlet macaws in the Amazon **Rain Forest**. They stand out!

Look! There is a toco toucan. Is that a guava fruit in its giant **bill**?

Gathered in Groups

Stay on the lookout for yellow birds as we head to central Africa. They are village weavers.

There is a group of them now! These busy birds build round, hanging nests.

Dusk is falling here in Europe. That **flock** of common starlings will swoop down soon.

These songbirds sleep in trees. Living in big groups keeps them safe from the tawny owl's **talons**!

tawny owl
talons

Roosts and Nests

Here in India, many birds make homes. Those gray herons are **roosting** after a long day of fishing.

That peacock will roost soon, too.
For now, it shows off for a **mate**!

Listen carefully as we explore New Zealand's forests. Laughing kookaburras call from their nests in tree holes.

Look low to spot the North Island brown kiwi. These flightless birds live in **burrows**!

JIMMY SAYS

The North Island brown kiwi comes out at night to hunt. Its long bill helps it pull worms from the ground!

Long Journeys

Here we are in icy Antarctica.
Those Arctic terns just arrived.
They finished their fall **migration**!

These emperor penguins are heading out. They will find food for their young.

JIMMY SAYS

Arctic terns have the longest migration of any animal on Earth. They fly about 25,000 miles (40,234 kilometers) each year!

Our trip around the world
is over. It is time to return home.

Next time you are outside,
look around. Busy birds are fun
to watch!

Bird Facts

Glossary

bill–the mouth of a bird

binoculars–a tool that people can look through to see far-away things up close

burrows–holes in the ground used as homes by some animals

carrion–the rotting meat of a dead animal

flock–a group of birds

hollow–empty through the middle

mate–a partner to produce offspring

migration–the act of traveling from one place to another, often with the seasons

nectar–a sweet liquid made by plants

rain forest–a thick, green forest that receives a lot of rain

roosting–settling down to sleep, often in a high place

talons–sharp claws on birds that allow them to grab and tear into their food

warm-blooded–having a high body temperature that does not change with the environment

To Learn More

AT THE LIBRARY

Editors of Storey Publishing. *Bird Watch: What Will You Find?* North Adams, Mass.: Storey Publishing, 2020.

Hansen, Grace. *Arctic Tern Migration.* Minneapolis, Minn.: Abdo Kids, 2021.

Riggs, Kate. *Hummingbirds.* Mankato, Minn.: The Creative Company, 2023.

ON THE WEB

FACTSURFER

Factsurfer.com gives you a safe, fun way to find more information.

1. Go to www.factsurfer.com.
2. Enter "brilliant birds" into the search box and click 🔍.
3. Select your book cover to see a list of related content.

BEYOND THE MISSION

> WHAT PLACE DID YOU THINK HAD THE MOST INTERESTING BIRDS? WHY?

> MAKE UP YOUR OWN BIRD. WHAT DOES IT LOOK LIKE? DRAW A PICTURE OF YOUR BIRD.

> CAN YOU THINK OF ANOTHER BIRD THAT WAS NOT IN THE BOOK?

Index